Contents

Introduction

If you are wild about learning and wild about animals – this book is for you! It will take you on a wild adventure, where you will practise key problem-solving and reasoning skills and explore the animals of the Serengeti along the way.

Each topic is introduced in a clear and simple way, with lots of interesting activities to complete, so that you can practise what you have learned.

You should attempt the tasks without the use of a calculator at first, but calculators may be used to check your answers.

Alongside every topic, you will uncover fascinating facts about the amazing animals found in the Serengeti ecosystem in Africa.

When you have completed each topic, record the animals that you have seen and the skills that you have learned in the explorer's logbook on pages 44–45.

Good luck, explorer! Don't get too close!

**Stephen Monaghan &
Melissa Blackwood**

Place value

Place value
is important because
it tells you the value of each
digit in a number.
For example, in the number:
H T U
2 4 5
The 2 is worth 200.
The 4 is worth 40.
The 5 is worth 5.

FACT FILE

Animal: Chameleon
Habitat: Forests. Madagascar contains half of the world's chameleon population
Weight: 0.5 to 408 g
Lifespan: 5 to 12 years
Diet: Insects, snails, spiders, lizards, birds, flowers, berries and fruit

Task 1 Write down the value (in numbers) of the underlined number of flies eaten by a chameleon.

a 6<u>7</u>6

The value of the 7 = _____

b 88<u>1</u>

The value of the 1 = _____

c <u>1</u>29

The value of the 1 = _____

d 9<u>8</u>3

The value of the 8 = _____

Task 2 Write down the following number of flies eaten in digits.

a Three hundred and forty-six _____

b Seven hundred and thirty _____

c Nine hundred and fifty-one _____

d One hundred and ninety-nine _____

Task 3 Scientists have put the number of flies eaten into an abacus.

WILD FACT

A chameleon can look at two things at the same time. This is because its eyes can rotate and swivel independently!

For example:

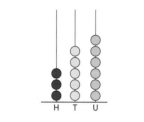

$100 + 50 + 8 = 158$

Work out how many flies were eaten.

a

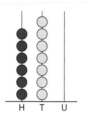

_____

b

_____

c

_____

d

_____

WILD FACT

Chameleons' tongues can be as long as their bodies!

Exploring Further ...

Draw an abacus to show the number of insects eaten by each chameleon.

a 718

b 168

Now crawl to pages 44–45 to record what you have learned in your explorer's logbook.

More place value

Place value is the value of each digit in a number. These can include units, tens, hundreds and thousands.

For example, in the number 453:

The place value of the 4 is '4 hundreds'.

The place value of the 5 is '5 tens'.

The place value of the 3 is '3 units'.

To find one more hundred we only need to change the hundred column:

$$
\begin{array}{r}
453 \\
+100 \\
\hline
553 \\
\hline
\end{array}
$$

To change the tens we only change the tens column:

$$
\begin{array}{r}
453 \\
+\,10 \\
\hline
463 \\
\hline
\end{array}
$$

Task 1 — Each pack of mongooses below adds 10 more. Work out the new total (remember the tens column will change).

a
$$
\begin{array}{r}
67 \\
+\,10 \\
\hline
 \\
\end{array}
$$

b
$$
\begin{array}{r}
45 \\
+\,10 \\
\hline
 \\
\end{array}
$$

c
$$
\begin{array}{r}
58 \\
+\,10 \\
\hline
 \\
\end{array}
$$

d
$$
\begin{array}{r}
24 \\
+\,10 \\
\hline
 \\
\end{array}
$$

e
$$
\begin{array}{r}
42 \\
+\,10 \\
\hline
 \\
\end{array}
$$

f
$$
\begin{array}{r}
33 \\
+\,10 \\
\hline
 \\
\end{array}
$$

Task 2   Work out the number of mongooses below if **100** is added to the pack.

a   120
    + 100
    _____

b   458
    + 100
    _____

c   234
    + 100
    _____

d   342
    + 100
    _____

e   432
    + 100
    _____

f   673
    + 100
    _____

WILD FACT

Some species of mongoose will daringly take on venomous snakes such as cobras.

WILD FACT

Mongooses are very social animals that live in big packs with up to 50 individuals.

Task 3   Now work out the number of mongooses if **100** is taken away.

a   120
    – 100
    _____

b   458
    – 100
    _____

c   234
    – 100
    _____

d   342
    – 100
    _____

e   432
    – 100
    _____

f   968
    – 100
    _____

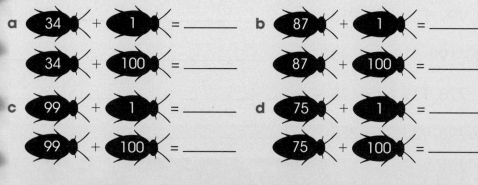

Exploring Further ...

Add the numbers on the bugs.

For example: 47 + 1 = 48

47 + 100 = 147

a 34 + 1 = _____

34 + 100 = _____

b 87 + 1 = _____

87 + 100 = _____

c 99 + 1 = _____

99 + 100 = _____

d 75 + 1 = _____

75 + 100 = _____

Now waddle to pages 44–45 to record what you have learned in your explorer's logbook.

Ordering numbers

FACT FILE

Animal: Leopard
Habitat: Bush and riverine forest
Weight: Up to 96 kg
Lifespan: 12 to 21 years
Diet: Leopards eat a wide variety of food, ranging from zebra to crab!

WILD FACT

Leopards can have as many as 1000 spots on their bodies!

Task 1

Scientists list the number of spots on wild leopards. Place the numbers in order from smallest to largest to help them.

a 674, 999, 728, 398, 918 _____

b 1000, 198, 718, 485, 482 _____

c 291, 728, 179, 685, 638 _____

d 956, 798, 938, 976, 838 _____

Task 2 Write the smallest and largest number you can make with the four digits provided.

a

| 7 | 2 | 9 | 2 |

Smallest = _____   Largest = _____

b

| 1 | 8 | 3 | 2 |

Smallest = _____   Largest = _____

c

| 1 | 6 | 9 | 2 |

Smallest = _____   Largest = _____

Task 3 Arrange the numbers from smallest to largest on the number line below.

For example: **456   171   782   891   719**

```
                              782
        171         456     719     891
    0           500             1000
```

a 647, 191, 829, 999   **b** 819, 157, 181, 918

```
    0                   1000        0                   1000
```

Exploring Further ...

Create the two largest and the two smallest numbers from these digits.

| 2 | 7 | 1 | 8 | _____ _____ _____ _____

Now place them in the correct order on the number line below.

```
    0                   10000
```

Now sprint to pages 44–45 to record what you have learned in your explorer's logbook.

Rounding numbers

You can **round numbers** up or down to the nearest 10, 100 or 1000. This can be a little difficult, so just remember:

Numbers 1, 2, 3, 4 – **round down**

Numbers 5, 6, 7, 8, 9 – **round up**

For example, 2456 rounded:

To the nearest **10**: 2460

To the nearest **100**: 2500

To the nearest **1000**: 2000

Task 1

Scientists have been counting the number of crested porcupines in a certain area. Round these numbers to the nearest 10.

a 57 _____

b 91 _____

c 99 _____

d 157 _____

e 82 _____

f 138 _____

g 37 _____

h 189 _____

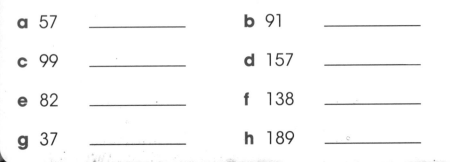

Task 2

Help the scientists record the number of females in the area to the nearest 100.

a 234 _____   b 789 _____

c 101 _____   d 199 _____

e 457 _____   f 849 _____

WILD FACT

It is called a 'crested porcupine' because it can raise its black and white quills along its head and back into a crest.

Task 3

The number of male crested porcupines varies from year to year. Help the scientists to round these numbers to the nearest 1000.

a 9866 _____   b 4571 _____

c 2234 _____   d 8192 _____

e 7191 _____   f 6799 _____

g 3488 _____   h 1167 _____

Exploring Further ...

Round the numbers in the apples to the nearest 10, 100 and 1000.

a 5728
10: _____
100: _____
1000: _____

b 6819
10: _____
100: _____
1000: _____

c 4311
10: _____
100: _____
1000: _____

d 9999
10: _____
100: _____
1000: _____

Now carefully crawl to pages 44–45 to record what you have learned in your explorer's logbook.

Roman numerals

Hundreds of years ago, when the Romans ruled, they used a number system called **numerals**. Today we call these **Roman numerals**. You may have seen them on a clock like this:

The Romans used the following symbols to represent numbers.

1	5	10	50	100	500	1000
I	V	X	L	C	D	M

To write different numbers, we combine the symbols.

- The same number is repeated no more than three times in a row.

- If a symbol representing a smaller value is written after a symbol representing an equal or larger number, we add the values together.

- If a symbol representing a smaller value is written before a symbol representing a larger value, we subtract the smaller number from the larger.

For example:

VI = 6 and IV = 4   LX = 60 and XL = 40

Task 1

Researchers have studied the wingspans of tawny eagles. Write the Roman numeral next to each number to complete the table.

Eagle	Wingspan (cm)	Roman numeral
Example	22	XXII
A	10	
B	32	
C	46	
D	50	
E	67	
F	72	

Task 2   Work out the numbers of tawny eagles.

For example: LXVII = L + X + V + I which is 50 + 10 + 5 + 1 = 66.

a XVII = _____   b XCV = _____

c LVII = _____   d XXXVIII = _____

e LXI = _____   f DCV = _____

WILD FACT

The tawny eagle attacks mammals as big as hares, large birds and lizards with one swoop and a grasp of its claws.

Task 3   Work out the following:

a XXII + XXXV = _____

b XXXI + LIV = _____

c LXIII + XXVI = _____

d LV + XXII = _____

e LXXI + XXVIII = _____

f LXV + XXXII = _____

g LVI + VII = _____

WILD FACT

The tawny eagle is about 62 to 72 cm long.

Exploring Further ...

Change the Roman numerals on the coins to numbers.

a (XXVII) = _____   b (LXVI) = _____

c (LXXIV) = _____   d (XL) = _____

Now fly to pages 44–45 to record what you have learned in your explorer's logbook.

Number sequences

A number sequence has a **rule**, which is a way to find the value of the next term. For example, in this sequence the rule is +4: 0, 4, 8, 12, 16, 20

FACT FILE

Animal: Bat-eared fox
Habitat: Short grass plains
Weight: 2.2 to 4.5 kg
Lifespan: 6 to 12 years
Diet: Termites, dung beetles and other insects

Task 1

Scientists have tracked some bat-eared foxes and have noticed specific patterns, which form number sequences. Write down the rule for each number sequence.

a 12, 16, 20, 24, 28, 32, 36   Rule = _____

b 16, 24, 32, 40, 48, 56, 64   Rule = _____

c 18, 27, 36, 45, 54, 63, 72   Rule = _____

d 7, 14, 21, 28, 35, 42, 49   Rule = _____

e 12, 18, 24, 30, 36, 42, 48   Rule = _____

WILD FACT

A bat-eared fox's ears can grow up to 13.4 cm long.

Task 2

Scientists have noticed the number of bat-eared foxes going down in the outer regions of the plains. Write down the next five numbers in the sequence if the trend continues.

a 78, 72, 66, 60, [ ] , [ ] , [ ] , [ ] , [ ]

b 72, 64, 56, 48, [ ] , [ ] , [ ] , [ ] , [ ]

c 48, 44, 40, 36, [ ] , [ ] , [ ] , [ ] , [ ]

d 99, 90, 81, 72, [ ] , [ ] , [ ] , [ ] , [ ]

e 70, 63, 56, 49, [ ] , [ ] , [ ] , [ ] , [ ]

Task 3

Now apply your knowledge of sequences to fill in the blanks.

a Rule: +4

10, 14, [ ] , 22, [ ] , 30, [ ]

b Rule: −6

55, 49, [ ] , 37, [ ] , 25, [ ]

c Rule: +9

6, 15, [ ] , 33, [ ] , 51, [ ]

d Rule: −8

50, 42, [ ] , 26, [ ] , 10, [ ]

WILD FACT

Bat-eared foxes use their amazing ears to detect their food. Their ears are so good that they can hear larvae chewing their way out of an underground dung beetle ball!

Exploring Further ...

Now use your knowledge to complete the sequences.

a Rule: +8

12, 20, [ ] , [ ] , [ ] , [ ]

b Rule: −7

47, 40, [ ] , [ ] , [ ] , [ ]

Now hunt your way to pages 44–45 to record what you have learned in your explorer's logbook.

13

More number sequences

FACT FILE

Animal:	African elephant
Habitat:	Dense forests and open plains
Weight:	African elephants are the earth's largest land animals weighing 2300 to 6300 kg
Lifespan:	70 years
Diet:	Plants of almost any size

Task 1 Scientists have noticed patterns in the number of litres of water drunk by some elephants. Predict the next three numbers (and their rules) in these sequences.

a. 50 , 75 , 100 , ___ , ___ , ___   Rule: _____

b. 50 , 100 , 150 , ___ , ___ , ___   Rule: _____

c. 150 , 300 , 450 , ___ , ___ , ___   Rule: _____

d. 125 , 150 , 175 , ___ , ___ , ___   Rule: _____

e. 10 , 60 , 110 , ___ , ___ , ___   Rule: _____

Task 2 Work out the rule of the following sequences, which show the number of elephants in certain areas of Africa.

WILD FACT
An elephant's trunk contains about 100 000 different muscles!

a 800, 700, 600, 500, 400, 300   Rule: _____

b 700, 675, 650, 625, 600, 575   Rule: _____

c 8250, 7250, 6250, 5250, 4250   Rule: _____

d 525, 475, 425, 375, 325, 275   Rule: _____

e 1340, 1240, 1140, 1040, 940   Rule: _____

Task 3 Scientists tracked the weight of some African elephants over three years. Predict the next two weights if this pattern continues and write down each rule.

a 230 g, 330 g, 430 g, ☐ g, ☐ g   Rule: _____

b 1200 g, 1300 g, 1400 g, ☐ g, ☐ g   Rule: _____

c 1250 g, 1300 g, 1350 g, ☐ g, ☐ g   Rule: _____

d 2345 g, 3345 g, 4345 g, ☐ g, ☐ g   Rule: _____

Exploring Further ...

Apply your number sequence knowledge to these problems.

a Rule: +1000

1800, 2800, 3800, ☐ , ☐ , ☐

b Rule: −50

375, 325, 275, ☐ , ☐ , ☐

c Rule: +25

135, 160, 185, ☐ , ☐ , ☐

d Rule: −100

2460, 2360, 2260, ☐ , ☐ , ☐

WILD FACT
An African elephant drinks 18 to 26 litres of water a day and can drink up to 152 litres a day in high temperatures.

Now stampede to pages 44–45 to record what you have learned in your explorer's logbook.

Adding four digit numbers

Adding four digit numbers can be tricky. So we can use the **columnar method** to make things a little easier.

$$3239 + 5882 = \boxed{}$$

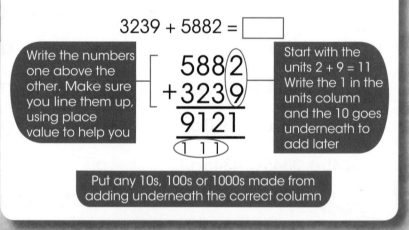

Write the numbers one above the other. Make sure you line them up, using place value to help you

```
  5882
+ 3239
  9121
  1 1 1
```

Start with the units 2 + 9 = 11 Write the 1 in the units column and the 10 goes underneath to add later

Put any 10s, 100s or 1000s made from adding underneath the correct column

Task 1

Scientists have been researching the weight of male and female pairs of giraffes. Calculate their combined weight (don't forget to put the unit of measurement – in this case, kg).

a
```
  1135 kg
+ 1245 kg
_____
```

b
```
  1363 kg
+ 1198 kg
_____
```

c
```
  1399 kg
+ 1037 kg
_____
```

d
```
  1310 kg
+  999 kg
_____
```

e
```
  1240 kg
+ 1270 kg
_____
```

f
```
  1020 kg
+  805 kg
_____
```

FACT FILE

Animal: Giraffe
Habitat: Savannahs of Africa
Weight: 794 to 1270 kg
Lifespan: 25 years
Diet: Trees and shrubs

Task 2

Use the column addition method to add up the total number of spots on each herd of giraffes.

a   6356
   + 3523
_____

b   6462
   + 3523
_____

c   3569
   + 4333
_____

d   3492
   + 2738
_____

e   4563
   + 3456
_____

f   7654
   + 1419
_____

Task 3

Use the column addition method to work out the lifespan of a male and female pair of giraffes in days.

a   5333
   + 4232
_____

b   2345
   + 4312
_____

c   2182
   + 5238
_____

d   4681
   + 3891
_____

e   3451
   + 5671
_____

f   6349
   + 3922
_____

WILD FACT

The spots on a giraffe can act as fantastic camouflage.

Exploring Further ...

Now use your column addition skills to find the total amount of water consumed each week by a male and female pair.

a   2310 litres
   +4502 litres
_____

b   6315 litres
   +2714 litres
_____

Now stretch to pages 44–45 to record what you have learned in your explorer's logbook.

Subtracting four digit numbers

When **subtracting four digit numbers**, you can also use the **columnar method**.

$$2416 - 387 = \boxed{}$$

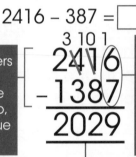

Write the numbers one above the other. Make sure you line them up, using place value to help you

Start with the units 6 – 7 = can't do! So, take a 10 from the tens column to turn 6 into 16, then take 7 away

Keep moving across from right to left, units to hundreds etc, until you have subtracted every column

WILD FACT

Warthogs wallow in water (and even mud) to cool down!

Task 1

Use the columnar method to work out the difference between the most bugs eaten and the fewest bugs eaten by a group of warthogs.

a
```
  4679
- 2465
_____
```

b
```
  7292
- 5272
_____
```

c
```
  7812
- 4602
_____
```

d
```
  5562
- 3241
_____
```

e
```
  6372
- 4151
_____
```

f
```
  8459
- 3359
_____
```

FACT FILE

Animal: Warthog
Habitat: Savannah
Weight: Females: 45 to 75 kg; males: 60 to 150 kg
Lifespan: 15 years
Diet: Grass, roots, berries and fruit

Task 2

Use the column subtraction method to work out how much clean water is left once a group of warthogs has bathed in different pools.

a 8292 litres

  – 5271 litres

  _____

b 7789 litres

  – 5852 litres

  _____

c 9821 litres

  – 6210 litres

  _____

d 9819 litres

  – 8829 litres

  _____

e 6728 litres

  – 3836 litres

  _____

WILD FACT

Even though warthogs look ferocious, they are actually herbivores and would rather run than fight.

Task 3

Now use your column subtraction skills to work how much further warthog A walked than warthog B each day.

a Monday

Warthog A	6732 m
Warthog B	4623 m

b Tuesday

Warthog A	7882 m
Warthog B	6982 m

Exploring Further ...

Look at the distance in metres the warthogs walked on Saturday.

Warthog A	7389 m
Warthog B	8272 m

a   Which warthog walked the furthest, A or B? _____

b   How much further did it walk? _____

Now snort your way to pages 44–45 to record what you have learned in your explorer's logbook.

Addition and subtraction problems

Addition and subtraction questions can be presented in lots of different ways. Remember to use column addition and column subtraction to help you answer the problems on these pages. Top tip: read the questions carefully to understand what operation to use.

FACT FILE

Animal:	Cheetah
Habitat:	Open savannahs
Weight:	21 to 72 kg
Lifespan:	10 to 20 years
Diet:	Gazelles, small antelopes and other small mammals

Task 1

Work out the total number of km for each question.

a During a week, a cheetah runs 7822 km. It then runs 7283 km the following week. How far does the cheetah run in total?

b A cheetah runs 9289 km in a week. A second cheetah runs 7821 km. Work out the difference between the two cheetahs' distances.

Task 2

Work out the weight of food each pack of cheetahs eats over two weeks.

a

Week 1	6272 kg
Week 2	9198 kg

b

Week 1	7292 kg
Week 2	1772 kg

c

Week 1	7383 kg
Week 2	6373 kg

d

Week 1	6482 kg
Week 2	9918 kg

WILD FACT

Cheetahs can accelerate from 0 to 100 km/h in under 3 seconds!

Task 3

Use the columnar method to work out the answers to the following problems.

a A male cheetah lives for 5739 days and his female partner lives for 3525 days. How much longer does the male cheetah live?

b Cheetah A lives for 6738 days and cheetah B lives for 6372 days. How many days do they live for altogether?

WILD FACT

One of the fastest land animals in the wor cheetahs can reac speeds of up to 113 km/h!

Exploring Further ...

A cheetah eats 2373 kg of food in January. It eats 4648 kg in February and 3637 kg in March. How much does it eat altogether?

Now hunt down to pages 44–45 to record what you have learned in your explorer's logbook.

Multiplication

Multiplying can be very tricky! So, when multiplying, use the number line method to **partition the larger numbers** to make things easier.

$27 \times 6 = \boxed{}$

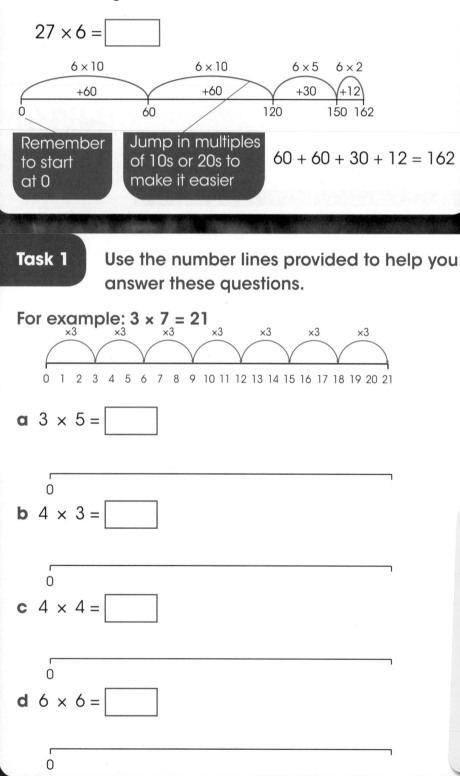

| 6 × 10 | 6 × 10 | 6 × 5 | 6 × 2 |
| +60 | +60 | +30 | +12 |

0   60   120   150 162

Remember to start at 0

Jump in multiples of 10s or 20s to make it easier

$60 + 60 + 30 + 12 = 162$

Task 1

Use the number lines provided to help you answer these questions.

For example: $3 \times 7 = 21$

×3 ×3 ×3 ×3 ×3 ×3 ×3

0 1 2 3 4 5 6 7 8 9 10 11 12 13 14 15 16 17 18 19 20 21

a $3 \times 5 = \boxed{}$

0

b $4 \times 3 = \boxed{}$

0

c $4 \times 4 = \boxed{}$

0

d $6 \times 6 = \boxed{}$

0

Task 2 Now use the blank number line below to answer the following questions.

WILD FACT

Lesser flamingos collect their food by holding their beak upside down in the water. This is called being a filter feeder.

a A flamingo spends 2 months in an area with its family. This is about 8 weeks. How many days is this? (Weeks × days = ?) [ ]

```
┌─────────────────────────────────┐
0
```

b A flamingo eats 8 crustaceans in a week. How many crustaceans will it eat in 5 weeks? [ ]

```
┌─────────────────────────────────┐
0
```

c A flamingo walks 9 km a week for 6 weeks. How far does it walk altogether? [ ] km

```
┌─────────────────────────────────┐
0
```

d A flamingo holds its beak under water for 9 seconds to get food. It does this 8 times. How long is this in total? [ ] seconds

WILD FACT

The lesser flamingo is the smallest species of flamingo. They only grow up to 1 metre in height.

```
┌─────────────────────────────────┐
0
```

Exploring Further ...

Use this number line to answer the following questions.

```
┌─────────────────────────────────┐
0
```

a Flamingo A eats 6 bugs every day. How many bugs will it eat in 7 days? [ ] bugs

b Flamingo B eats 9 bugs every day. How many bugs will it eat in 6 days? [ ] bugs

c Flamingo C eats 12 bugs every day. How many bugs will it eat in 4 days? [ ] bugs

d Which flamingo ate the most bugs? _____

Now gracefully walk to pages 44–45 to record what you have learned in your explorer's logbook.

To make dividing easier you can **use a number line** to help you.

$96 \div 8 = \boxed{}$

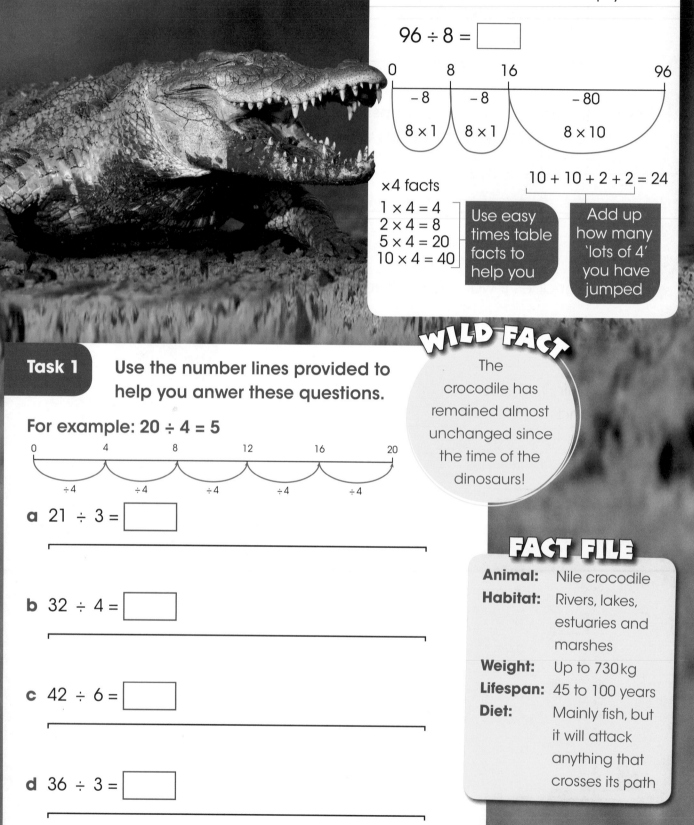

0   8   16   96

− 8   − 8   − 80

8 × 1   8 × 1   8 × 10

$10 + 10 + 2 + 2 = 24$

×4 facts

$1 \times 4 = 4$
$2 \times 4 = 8$
$5 \times 4 = 20$
$10 \times 4 = 40$

Use easy times table facts to help you

Add up how many 'lots of 4' you have jumped

WILD FACT

The crocodile has remained almost unchanged since the time of the dinosaurs!

Task 1

Use the number lines provided to help you anwer these questions.

For example: $20 \div 4 = 5$

0   4   8   12   16   20

÷ 4   ÷ 4   ÷ 4   ÷ 4   ÷ 4

a $21 \div 3 = \boxed{}$

b $32 \div 4 = \boxed{}$

c $42 \div 6 = \boxed{}$

d $36 \div 3 = \boxed{}$

FACT FILE

Animal:	Nile crocodile
Habitat:	Rivers, lakes, estuaries and marshes
Weight:	Up to 730 kg
Lifespan:	45 to 100 years
Diet:	Mainly fish, but it will attack anything that crosses its path

Task 2 Use the blank number lines below to work out the average weight of 1 crocodile (the total weight of crocodiles divided by how many there are).

a The total weight of 8 young crocodiles is 64 kg. How much does 1 weigh?

[ ] kg

b The total weight of 7 young crocodiles is 49 kg. How much does 1 weigh?

[ ] kg

c The total weight of 9 young crocodiles is 45 kg. How much does 1 weigh?

[ ] kg

d The total weight of 9 young crocodiles is 63 kg. How much does 1 weigh?

[ ] kg

WILD FACT

A large crocodile can easily kill and chomp down a wildebeest, zebra and even Cape buffalo!

Exploring Further ...

Work out how far each crocodile swam on average each day (total km swum divided by number of days).

a Crocodile A swam 48 km in 12 days.

_____

b Crocodile B swam 55 km in 11 days.

_____

c Crocodile C swam 63 km in 9 days.

_____

d Which crocodile swam the furthest in 1 day? _____

Now snap to pages 44–45 to record what you have learned in your explorer's logbook.

25

Mixed multiplication and division

Use your knowledge of **multiplication** and **division** to answer the problems on this page. Remember you can use a number line for multiplication and division.

Task 1 Use the number lines provided to help you answer these questions. Top tip: pay attention to the symbol used.

For example: 5 × 6 = 30

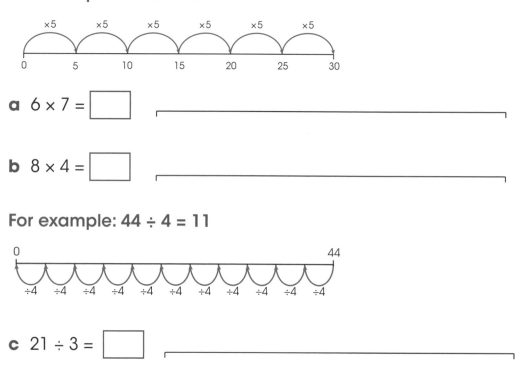

a 6 × 7 = ☐

b 8 × 4 = ☐

For example: 44 ÷ 4 = 11

c 21 ÷ 3 = ☐

Task 2

There are 7 female colobus monkeys for every male. Work out how many females there are below.

For example: there are 4 males.
So 4 × 7 = 28

0   4   8   12   16   20   24   28

a There are 6 males. How many females are there?

b There are 8 males. How many females are there?

c There are 4 males. How many females are there?

d There are 3 males. How many females are there?

WILD FACT

The name 'colobus' comes from the Greek word meaning 'mutilated', because colobus monkeys do not have any thumbs!

FACT FILE

Animal: Colobus monkey
Habitat: Forest
Weight: 7.8 to 13.5 kg
Lifespan: Up to 20 years
Diet: Mainly leaves, flowers and unripe fruit

WILD FACT

Black and white colobus monkeys talk loudly to other tribe members, either to mark territory or to warn members of danger.

Exploring Further ...

Work out the following using the number line below for help.

0   60

a 48 ÷ 12 = _____   **b** 54 ÷ 6 = _____

c 11 × 4 = _____   **d** 7 × 7 = _____

Now swing to pages 44–45 to record what you have learned in your explorer's logbook.

Fractions

Fractions can be really tricky to read, but we can break them down simply.

For example, if 2 children out of 4 like football, you can write it like this:

The top number tells you how many you have

The bottom number tells you how many there are all together

We call the top number the **numerator**: the number of something you have.

We call the bottom number the **denominator**: a whole number that is split into equal parts.

$$\frac{\text{Numerator}}{\text{Denominator}}$$

If you forget these names, just think '**Down'ominator**'.

Task 1

Look at the fractions and write down which animal the hyena ate more of.

a warthogs: $\frac{2}{5}$ zebras: $\frac{3}{5}$

Hyena A ate more _____

b warthogs: $\frac{2}{6}$ zebras: $\frac{4}{6}$

Hyena B ate more _____

c warthogs: $\frac{3}{4}$ zebras: $\frac{1}{4}$

Hyena C ate more _____

FACT FILE

Animal: Spotted hyena

Habitat: Savannahs, woodlands, grasslands, sub-deserts and mountains

Weight: 50 to 86 kg

Lifespan: Up to 25 years

Diet: Antelope, zebra, warthog, Cape buffalo, giraffe, hippopotamus and rhinoceros

Task 2

Complete the tables. Order them by most eaten (1st) to least eaten (4th). The first part has been done for you.

Hyena	Cape buffalo eaten	Fraction eaten	Order
A	5	$\frac{5}{13}$	1st
B	4	$\frac{}{13}$	
C	1	$\frac{}{13}$	
D	3	$\frac{}{13}$	

Hyena	Antelope eaten	Fraction eaten	Order
A	2	$\frac{2}{10}$	3rd
B	4		
C	1		
D	3		

Task 3

Now answer these questions.

a Which hyena from Task 2 ate the most Cape buffalo? _____

b Which hyena from Task 2 ate the fewest Cape buffalo? _____

Give your answers in tenths (e.g. $\frac{4}{10}$)

c How many antelope did hyenas A and B eat altogether?

d How many more antelope did hyena D eat than hyena C?

Exploring Further ...

Now put these fractions in order from smallest to largest:

$$\frac{2}{6}, \frac{3}{6}, \frac{5}{6}, \frac{1}{6}, \frac{6}{6}, \frac{4}{6}$$

Now laugh your way to pages 44–45 to record what you have learned in your explorer's logbook.

Decimal numbers

Ordering decimals simply means putting them in order from smallest to largest or from largest to smallest. Write down the numbers in a **place value chart** and compare the digits in each column, starting on the left.

For example, to compare 3.34 to 3.56:

H T U . t h
   ③ . ③ 4
   ③ . ⑤ 6

The units are both 3

But the tenths are different. 5 tenths is larger than 3 tenths so 3.56 is the largest number

Task 1   Work out on which day the jackals ate the most insects.

a Monday = 4.45   Tuesday = 8.67
H   T   U   .   t   h

The jackals ate the most insects on _____

b Wednesday = 6.66   Thursday = 6.67
H   T   U   .   t   h

The jackals ate the most insects on _____

c Sunday = 5.34   Monday = 5.43
H   T   U   .   t   h

The jackals ate the most insects on _____

Task 2

Match the average speed (the decimal number) to the nearest whole number.

a 12.34 km/h   9 km/h

b 8.98 km/h   16 km/h

c 7.71 km/h   12 km/h

d 14.32 km/h   14 km/h

e 15.89 km/h   8 km/h

Task 3

Round the number of berries eaten by a pack of jackals to the nearest whole number.

a 10.45 = _____   b 5.78 = _____

c 16.50 = _____   d 2.22 = _____

Exploring Further ...

Work out which number is bigger.

a   4.56 or 4.76   The bigger number is _____

b   6.79 or 6.77   The bigger number is _____

Round these numbers to the nearest whole number.

c   4.78 to the nearest whole number = _____

d   11.03 to the nearest whole number = _____

Now run sharply to pages 44–45 to record what you have learned in your explorer's logbook.

31

Fractions and percentages

FACT FILE

Animal:	Crowned crane
Habitat:	Wetlands
Weight:	Up to 3.5 kg
Lifespan:	Over 30 years
Diet:	Plants, eggs, seeds, insects, frogs, worms, snakes, small fish and mammals

Fractions and percentages are just different ways of showing the **same value**.

For example, if $\frac{2}{4}$ animals are carnivores this means 50% are carnivores.

Here is a handy conversion chart. It will help you with the rest of the problems on this page.

Fraction	Percentage
$\frac{1}{4}$	25%
$\frac{1}{2}$	50%
$\frac{3}{4}$	75%
1	100%

Task 1 Match up the percentage to the correct fraction.

Task 2

Look at the fractions of the crowned crane population that visited the river over a four-week period. Convert them into percentages.

a $\frac{1}{2}$ = _____

b $\frac{3}{4}$ = _____

c 1 = _____

d $\frac{1}{4}$ = _____

WILD FACT

Crowned cranes have a rasping groan and a trumpeting call: 'u-wang u-wang'.

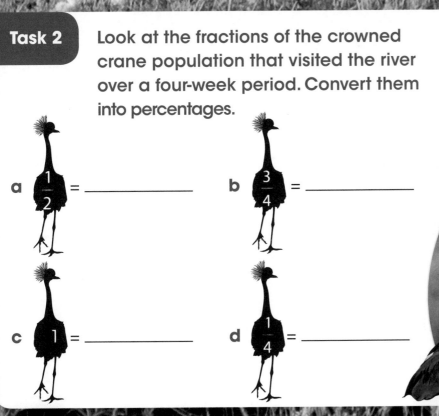

Task 3

Write down the missing fractions and percentages in the table so that scientists can compare them.

Crowned crane	Percentage of insects eaten in a week	Fraction of insects eaten in a week
A	25%	
B		$\frac{3}{4}$
C	100%	
D		$\frac{1}{2}$

WILD FACT

Unlike other cranes, crowned cranes bond and live in a pair all year round.

Exploring Further ...

Write down which crowned crane ate the most in Task 3. Rank them first (most), second, third and fourth (least).

First: _____

Second: _____

Third: _____

Fourth: _____

Now swoop to pages 44–45 to record what you have learned in your explorer's logbook.

Measuring and comparing

There are many different kinds of scales. Remember, when reading scales, pay attention to the increase of measure between each line.

For example, this scale goes up in increments of 100 g:

Task 1   Identify the weight of the aardvarks below.

a 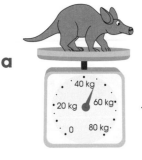 _____

b _____

c 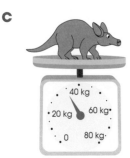 _____

d  _____

Task 2

Place the following aardvarks in order from tallest to smallest.

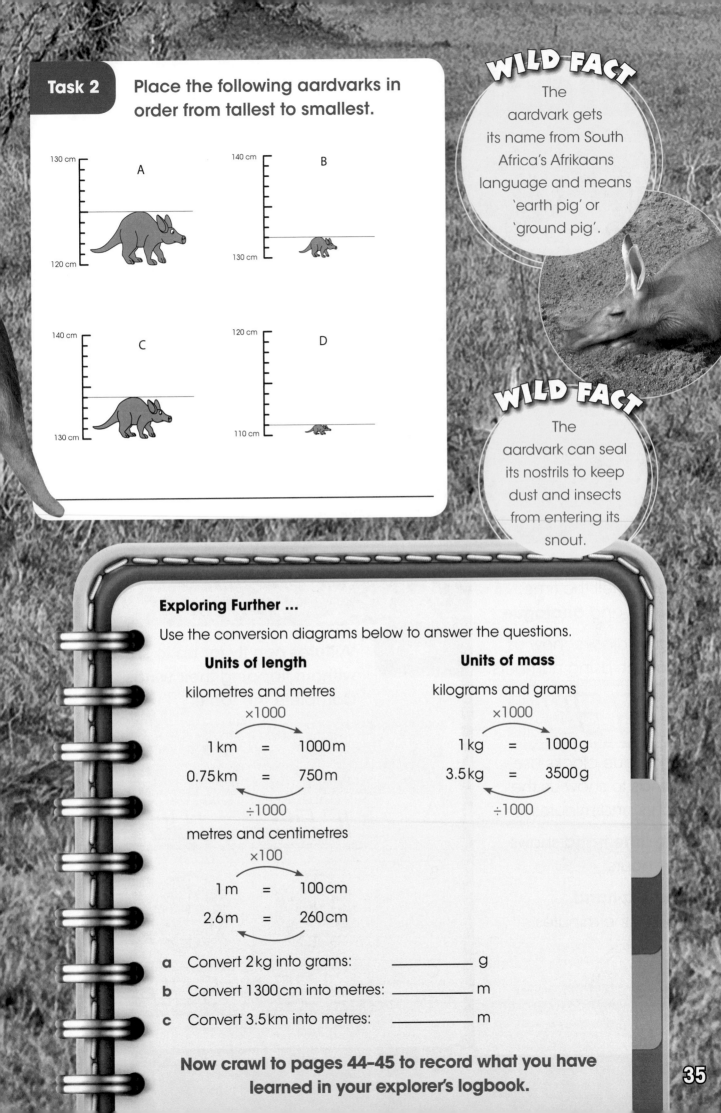

A — 130 cm / 120 cm

B — 140 cm / 130 cm

C — 140 cm / 130 cm

D — 120 cm / 110 cm

WILD FACT

The aardvark gets its name from South Africa's Afrikaans language and means 'earth pig' or 'ground pig'.

WILD FACT

The aardvark can seal its nostrils to keep dust and insects from entering its snout.

Exploring Further ...

Use the conversion diagrams below to answer the questions.

Units of length

kilometres and metres

×1000

1 km = 1000 m

0.75 km = 750 m

÷1000

metres and centimetres

×100

1 m = 100 cm

2.6 m = 260 cm

Units of mass

kilograms and grams

×1000

1 kg = 1000 g

3.5 kg = 3500 g

÷1000

a Convert 2 kg into grams: _____ g

b Convert 1300 cm into metres: _____ m

c Convert 3.5 km into metres: _____ m

Now crawl to pages 44–45 to record what you have learned in your explorer's logbook.

Calculating time

There are two different ways to tell the time: **digital** and **analogue**.

Digital clocks show us the time using numbers.

Analogue clocks use hands to show us the hours and minutes.

The **little hand** shows the hours.

The **big hand** shows the minutes.

Task 1 Vultures can fly for up to six hours without flapping their wings. Complete the table.

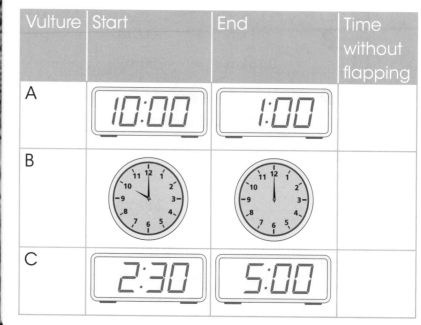

Vulture	Start	End	Time without flapping
A	10:00	1:00	
B	*(clock showing 11:00)*	*(clock showing 12:00)*	
C	2:30	5:00	

Task 2 Answer the following questions.

a Which of the vultures (A, B or C) in Task 1 went the longest without

flapping their wings? _____

b Vulture D did not flap its wings for 6 hours. It started this at 12.15.
Complete the analogue clocks to show the start time and finish time.

Start: 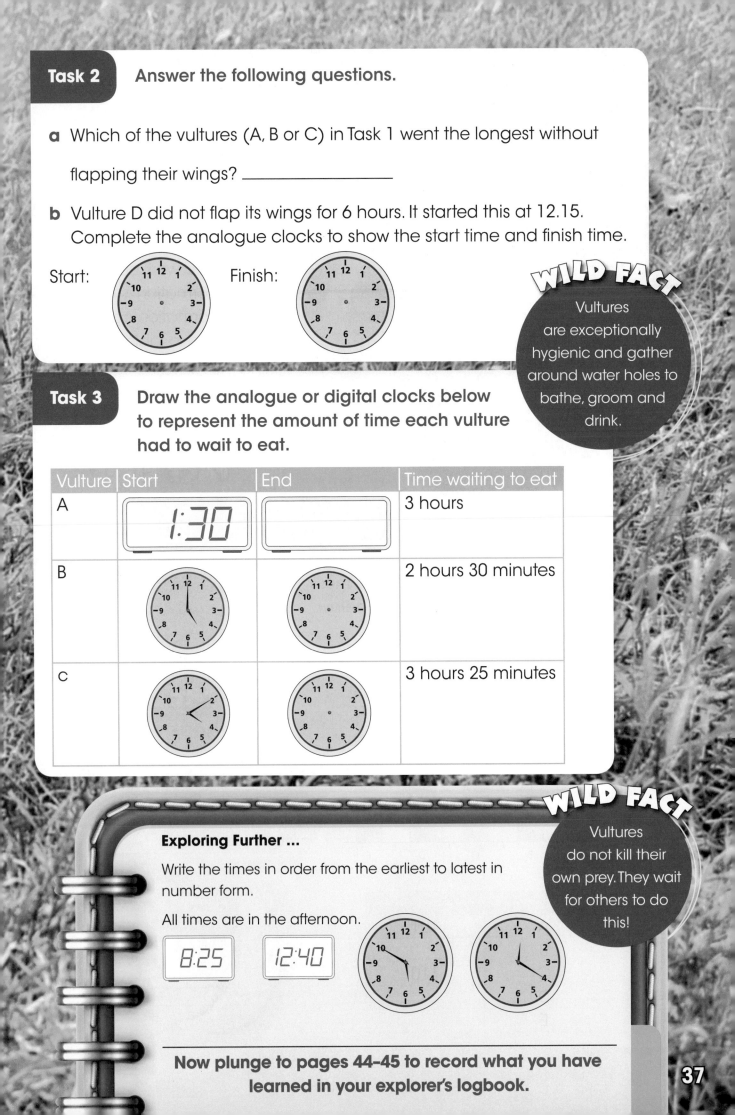   Finish:

Task 3 Draw the analogue or digital clocks below to represent the amount of time each vulture had to wait to eat.

Vulture	Start	End	Time waiting to eat
A	1:30		3 hours
B			2 hours 30 minutes
C			3 hours 25 minutes

Exploring Further ...

Write the times in order from the earliest to latest in number form.

All times are in the afternoon.

8:25   12:40

Now plunge to pages 44–45 to record what you have learned in your explorer's logbook.

Data problems

Data means information. To understand data we just need work out what the information is telling us. Information can be presented in tables, charts or graphs to make it easier to read.

FACT FILE

Animal: Ostrich
Habitat: Semi-arid plains and woodlands
Weight: 100 to 160 kg
Lifespan: 40 to 45 years
Diet: Plants and small invertebrates

WILD FACT

The long legs of an ostrich can be tough weapons. They are capable of killing a human or even a lion with their kick!

Task 1 Look at the chart and fill in the missing data in the table below.

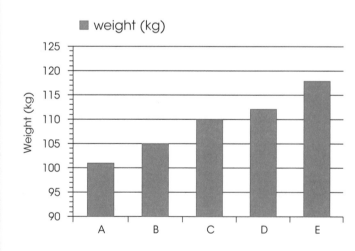

weight (kg)

Ostrich	Weight (kg)
A	101
B	
C	110
D	
E	

Task 2

Plot the data from the table on to the chart by drawing coloured bars for each ostrich.

Ostrich	Weight (kg)
A	120
B	103
C	114
D	119
E	107

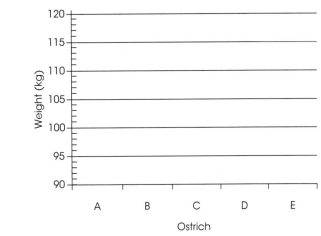

Task 3

Scientists measured how far four ostriches ran in a 10-minute period.

Look at the chart below and answer the following questions.

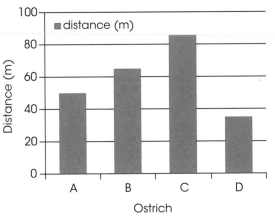

a Which ostrich ran the furthest distance? _____

b Which ostrich (A, B or C) ran the shortest distance?

_____

c Ostrich D ran 15 m less than ostrich A. Show this on the table.

Ostrich	Distance (m)
A	50
B	65
C	84
D	

WILD FACT

In certain African countries, people race each other on the backs of ostriches.

Exploring Further ...

Look at the table in task 3 and answer the following.

a What was the average distance travelled by all ostriches?
(The mean) _____

b What was the difference between the highest and the lowest?
(The range) _____

Now dash to pages 44–45 to record what you have learned in your explorer's logbook.

Area and perimeter

To find the **area** of a shape you can count the squares.

In this example each square represents 1 m.

1	2	3	4	5
6	7	8	9	10
11	12	13	14	15

So the area of this shape is 15 square metres or 15 m^2.

In maths we abbreviate 'square metres' to m^2. The 2 means 'squared'.

Perimeter is the total distance around the outer edge of a two-dimensional (2D) shape

This shape has a perimeter of 16 m, as 5 + 5 + 3 + 3 = 16

5 m

3 m   3 m

5 m

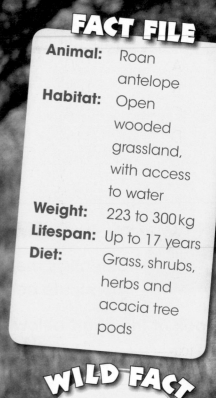
Task 1   Calculate how far the roan antelope roams during the day by working out the area covered.

Each square = 1 m

a   _____

b   _____

c   _____

d   _____

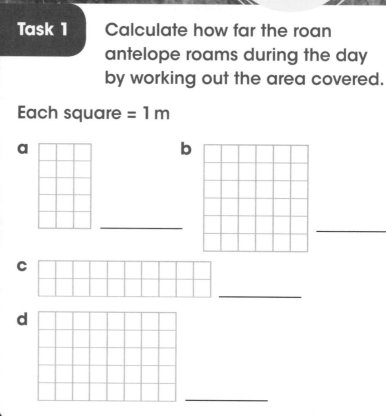

Task 2   Now try to work out the areas of these trickier shapes.

Each square = 1 m

a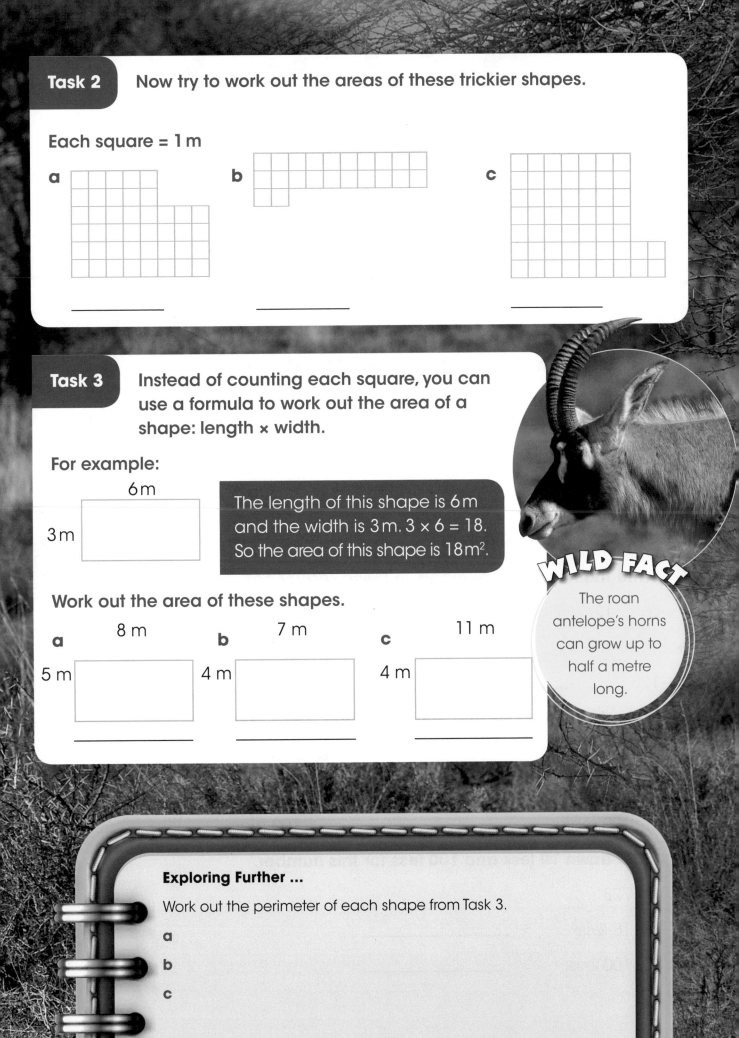

b

c

_____   _____   _____

Task 3   Instead of counting each square, you can use a formula to work out the area of a shape: length × width.

For example:

6 m

3 m

The length of this shape is 6 m and the width is 3 m. 3 × 6 = 18. So the area of this shape is 18 m².

Work out the area of these shapes.

a   8 m

5 m

b   7 m

4 m

c   11 m

4 m

_____   _____   _____

WILD FACT

The roan antelope's horns can grow up to half a metre long.

Exploring Further ...

Work out the perimeter of each shape from Task 3.

a

b

c

Now gallop to pages 44–45 to record what you have learned in your explorer's logbook.

Quick test

Now try these questions. Give yourself 1 mark for every correct answer – but only if you answer each part of the question correctly.

1 Write down the value of each digit in **3452.**

3 = _____   5 = _____

4 = _____   2 = _____

2 Order these numbers from largest to smallest: **3822, 6281, 348, 8891.**

_____

3 Round these numbers to the nearest **10, 100 and 1000.**

   a 3456

      10: _____

      100: _____

      1000: _____

   b 2787

      10: _____

      100: _____

      1000: _____

4 What number is this? **XVII** _____

5 Write the next three numbers in these sequences.

   a 8, 12, 16, ☐, ☐, ☐   **b** 28, 35, 42, ☐, ☐, ☐

6 Write the missing numbers in this sequence.

125, ☐, ☐, 200, ☐, ☐, 275

7 Write down 10 more and 100 more than these numbers.

   a 365

      10 more: _____

      100 more: _____

   b 618

      10 more: _____

      100 more: _____

8 Write down 10 less and 100 less for this number.

   728

      10 less: _____

      100 less: _____

9   2672
     $+3425$
    ___

10 7829
     -4529
    ___

11 There were 8373 bugs eaten in January and 3749 bugs eaten in February. How many were eaten altogether? Use column addition to answer this.

12 a $3 \times 12 =$ ☐   **b** $4 \times 7 =$ ☐   **c** $6 \times 4 =$ ☐

13 a $33 \div 3 =$ ☐   **b** $36 \div 4 =$ ☐   **c** $54 \div 6 =$ ☐

14 A colobus monkey eats 8 bugs each day. How many bugs does he eat in 3 days? _____

15 Circle the largest fractions.

$\frac{2}{5}$ or $\frac{3}{5}$ $\frac{2}{3}$ or $\frac{3}{3}$

16 Colour the matching fraction and percentage using the same colour.

(25%) (50%) (100%) (1) ($\frac{1}{4}$) ($\frac{1}{2}$)

17 Order these decimals from largest to smallest: **2.34 4.56 1.67 7.98**

_____

18 Round the following to the nearest whole number.

2.35 rounded: _____   10.9 rounded: _____

19 If a vulture starts to fly at (clock) and lands at (clock), how long is it in the air for? _____

20 Calculate the area of this shape.
Each square = 1 m

_____

43

Explorer's Logbook

Tick off the topics as you complete them and then colour in the star.

How do you feel?
- Needs practice
- Nearly there
- Got it!

Rounding numbers ☐

Adding four digit numbers ☐

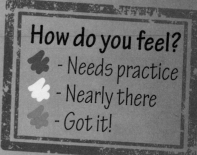
Roman numerals ☐

More number sequences ☐

Subtracting four digit numbers ☐

More place value ☐

Place value ☐

Ordering numbers ☐

Number sequences ☐

Decimal numbers ☐

Area and perimeter ☐

Division ☐

Fractions and percentages ☐

Addition and subtraction problems ☐

Multiplication ☐

Fractions ☐

Measuring and comparing ☐

Mixed multiplication and division ☐

Data problems ☐

Calculating time ☐

Answers

Pages 2–3
Task 1
a 70 **b** 1 **c** 100 **d** 80
Task 2
a 346 **b** 730 **c** 951 **d** 199
Task 3
a 300 + 50 + 6 = 356 **b** 900 + 80 + 7 = 987
c 600 + 70 = 670 **d** 600 + 10 + 8 = 618
Exploring Further...
a **b**

Pages 4–5
Task 1
a 77 **b** 55 **c** 68 **d** 34 **e** 52 **f** 43
Task 2
a 220 **b** 558 **c** 334 **d** 442 **e** 532 **f** 773
Task 3
a 20 **b** 358 **c** 134 **d** 242 **e** 332 **f** 868
Exploring Further...
a 34 + 1 = 35 **b** 87 + 1 = 88
34 + 100 = 134 87 + 100 = 187
c 99 + 1 = 100 **d** 75 + 1 = 76
99 + 100 = 199 75 + 100 = 175

Pages 6–7
Task 1
a 398, 674, 728, 918, 999 **b** 198, 482, 485, 718, 1000
c 179, 291, 638, 685, 728 **d** 798, 838, 938, 956, 976
Task 2
a Smallest = 2279 Largest = 9722
b Smallest = 1238 Largest = 8321
c Smallest = 1269 Largest = 9621
Task 3
a

b

Exploring Further...
Smallest: 1278, 1287
Largest: 8721, 8712

Pages 8–9
Task 1
a 60 **b** 90 **c** 100 **d** 160
e 80 **f** 140 **g** 40 **h** 190

Task 2
a 200 **b** 800 **c** 100 **d** 200 **e** 500 **f** 800
Task 3
a 10 000 **b** 5000 **c** 2000 **d** 8000
e 7000 **f** 7000 **g** 3000 **h** 1000
Exploring Further...
a 10: 5730 **b** 10: 6820
100: 5700 100: 6800
1000: 6000 1000: 7000
c 10: 4310 **d** 10: 10 000
100: 4300 100: 10 000
1000: 4000 1000: 10 000

Pages 10–11
Task 1
A = X B = XXXII C = XLVI
D = L E = LXVII F = LXXII
Task 2
a 17 **b** 95 **c** 57 **d** 38 **e** 61 **f** 605
Task 3
a 57 **b** 85 **c** 89 **d** 77 **e** 99 **f** 97 **g** 73
Exploring Further...
a 27 **b** 66 **c** 74 **d** 40

Pages 12–13
Task 1
a +4 **b** +8 **c** +9 **d** +7 **e** +6
Task 2
a 54, 48, 42, 36, 30 **b** 40, 32, 24, 16, 8
c 32, 28, 24, 20, 16 **d** 63, 54, 45, 36, 27
e 42, 35, 28, 21, 14
Task 3
a 10, 14, 18, 22, 26 30, 34 **b** 55, 49, 43, 37, 31, 25, 19
c 6, 15, 24, 33, 42, 51, 60 **d** 50, 42, 34, 26, 18, 10, 2
Exploring Further...
a 12, 20, 28, 36, 44, 52 **b** 47, 40, 33, 26, 19, 12

Pages 14–15
Task 1
a 50, 75, 100, 125, 150, 175 +25
b 50, 100, 150, 200, 250, 300 +50
c 150, 300, 450, 600, 750, 900 +150
d 125, 150, 175, 200, 225, 250 +25
e 10, 60, 110, 160, 210, 260 +50
Task 2
a –100 **b** –25 **c** –1000 **d** –50 **e** –100
Task 3
a 230 g, 330 g, 430 g, 530 g, 630 g +100
b 1200 g, 1300 g, 1400 g, 1500 g, 1600 g +100
c 1250 g, 1300 g, 1350 g, 1400 g, 1450 g +50
d 2345 g, 3345 g, 4345 g, 5345 g, 6345 g +1000

Exploring Further...
a 1800, 2800, 3800, 4800, 5800, 6800
b 375, 325, 275, 225, 175, 125
c 135, 160, 185, 210, 235, 260
d 2460, 2360, 2260, 2160, 2060, 1960

Pages 16–17
Task 1
a 2380 kg   **b** 2561 kg   **c** 2436 kg
d 2309 kg   **e** 2510 kg   **f** 1825 kg
Task 2
a 9879   **b** 9985   **c** 7902
d 6230   **e** 8019   **f** 9073
Task 3
a 9565   **b** 6657   **c** 7420   **d** 8572   **e** 9122   **f** 10271
Exploring Further...
a 6812 litres   **b** 9029 litres

Pages 18–19
Task 1
a 2214   **b** 2020   **c** 3210
d 2321   **e** 2221   **f** 5100
Task 2
a 3021 litres   **b** 1937 litres   **c** 3611 litres
d 990 litres   **e** 2892 litres
Task 3
a 2109 m   **b** 900 m
Exploring Further...
a Warthog B   **b** 883 m

Pages 20–21
Task 1
a 15 105 km   **b** 1468 km
Task 2
a 15 470 kg   **b** 9064 kg   **c** 13 756 kg   **d** 16 400 kg
Task 3
a 2214 days   **b** 13 110 days
Exploring Further...
10 658 kg

Pages 22–23
Task 1
a
0 3 6 9 12 15
Answer 15
b
0 4 8 12
Answer 12
c
0 4 8 12 16
Answer 16
d
0 6 12 18 24 30 36
Answer 36
Task 2
a
0 7 14 21 28 35 42 49 56
56 days

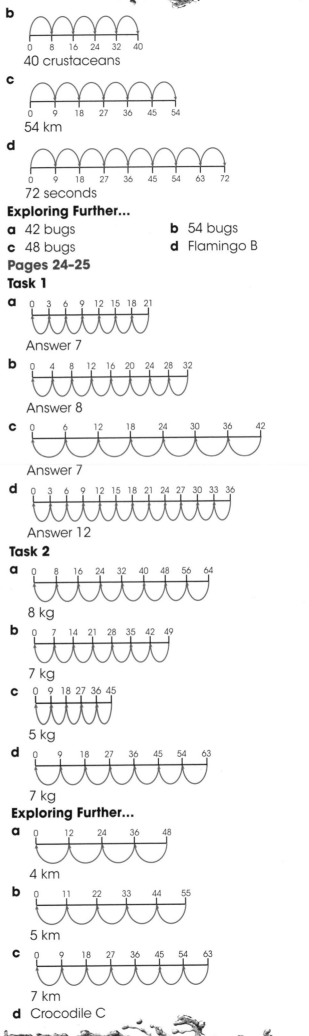

b
0 8 16 24 32 40
40 crustaceans
c
0 9 18 27 36 45 54
54 km
d
0 9 18 27 36 45 54 63 72
72 seconds
Exploring Further...
a 42 bugs   **b** 54 bugs
c 48 bugs   **d** Flamingo B

Pages 24–25
Task 1
a
0 3 6 9 12 15 18 21
Answer 7
b
0 4 8 12 16 20 24 28 32
Answer 8
c
0 6 12 18 24 30 36 42
Answer 7
d
0 3 6 9 12 15 18 21 24 27 30 33 36
Answer 12
Task 2
a
0 8 16 24 32 40 48 56 64
8 kg
b
0 7 14 21 28 35 42 49
7 kg
c
0 9 18 27 36 45
5 kg
d
0 9 18 27 36 45 54 63
7 kg
Exploring Further...
a
0 12 24 36 48
4 km
b
0 11 22 33 44 55
5 km
c
0 9 18 27 36 45 54 63
7 km
d Crocodile C

Pages 26–27

Task 1

a

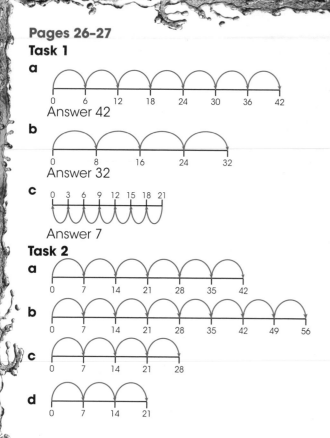

Answer 42

b

Answer 32

c

Answer 7

Task 2

a

b

c

d

Exploring Further...

a 4   **b** 9   **c** 44   **d** 49

Pages 28–29

Task 1

a zebras   **b** zebras   **c** warthogs

Task 2

Hyena	Cape buffalo eaten	Fraction eaten	Order
A	5	$\frac{5}{13}$	1st
B	4	$\frac{4}{13}$	2nd
C	1	$\frac{1}{13}$	4th
D	3	$\frac{3}{13}$	3rd

Hyena	Antelope eaten	Fraction eaten	Order
A	2	$\frac{2}{10}$	3rd
B	4	$\frac{4}{10}$	1st
C	1	$\frac{1}{10}$	4th
D	3	$\frac{3}{10}$	2nd

Task 3

a Hyena A   **b** Hyena C

c $\frac{4}{10} + \frac{2}{10} = \frac{6}{10}$ **d** $\frac{3}{10} - \frac{1}{10} = \frac{2}{10}$

Exploring Further...

$\frac{1}{6}, \frac{2}{6}, \frac{3}{6}, \frac{4}{6}, \frac{5}{6}, \frac{6}{6}$

Pages 30–31

Task 1

a Tuesday   **b** Thursday   **c** Monday

Task 2

12.34 km/h = 12 km/h
8.98 km/h = 9 km/h
7.71 km/h = 8 km/h
14.32 km/h = 14 km/h
15.89 km/h = 16 km/h

Task 3

a 10   **b** 6   **c** 17   **d** 2

Exploring Further...

a 4.76   **b** 6.79   **c** 5   **d** 11

Pages 32–33

Task 1

25% = $\frac{1}{4}$ 75% = $\frac{3}{4}$

50% = $\frac{1}{2}$ 100% = 1

Task 2

a 50%   **b** 75%   **c** 100%   **d** 25%

Task 3

Crowned crane	Percentage of insects eaten	Fraction of insects eaten
A	25%	$\frac{1}{4}$
B	75%	$\frac{3}{4}$
C	100%	1
D	50%	$\frac{1}{2}$

Exploring Further...

First: Crane C   Second: Crane B
Third: Crane D   Fourth: Crane A

Pages 34–35

Task 1

a 45 kg   **b** 50 kg   **c** 35 kg   **d** 60 kg

Task 2

C, B, A, D

Exploring Further...

a 2000 g   **b** 1.3 m   **c** 3500 m

Pages 36–37

Task 1

A 3 hours   **B** 2 hours
C 2 hours 30 minutes

Task 2

a Vulture A

b Start: Finish: